UFAD O&M GUIDE

This publication was developed under the auspices of ASHRAE Technical Committee (TC) 5.3, Room Air Distribution.For more information on ASHRAE TC 5.3, visit https://tc0503.ashraetcs.org/.

Updates and errata for this publication will be posted on the ASHRAE website at www.ashrae.org/publicationupdates.

UFAD O&M GUIDE

A Practical Guide for Operation and Maintenance of Underfloor Air Distribution Systems

ASHRAE
Atlanta

ISBN 978-1-939200-52-5 (paperback)
ISBN 978-1-939200-53-2 (PDF)

1791 Tullie Circle, NE • Atlanta, GA 30329 • www.ashrae.org
ASHRAE is a registered trademark of the American Society of Heating, Refrigerating and Air-Conditioning Engineers, Inc.

Printed in the United States of America
Cover design by Joseph Hullebusch

Library of Congress Cataloging-in-Publication Data

Names: American Society of Heating, Refrigerating and Air-Conditioning Engineers, author.
Title: UFAD O&M guide : a practical guide for operation and maintenance of underfloor air distribution systems.
Other titles: Practical guide for operation and maintenance of underfloor air distribution systems
Description: Atlanta, GA : ASHRAE, [2016]
Identifiers: LCCN 2016052059| ISBN 9781939200525 (pbk.) | ISBN 9781939200532 (PDF)
Subjects: LCSH: Air ducts--Maintenance and repair--Handbooks, manuals, etc. | Air conditioning--Maintenance and repair--Handbooks, manuals, etc. | Ventilation--Maintenance and repair--Handbooks, manuals, etc.
Classification: LCC TH7683.D8 U43 2016 | DDC 697.9/2--dc23 LC record available at https://lccn.loc.gov/2016052059

Contents

Acknowledgments

UFAD O&M Guide—A Practical Guide for Operation and Maintenance of Underfloor Air Distribution Systems was written to serve as a supplement to the *UFAD Guide—Design, Construction and Operation of Underfloor Air Distribution Systems* published in 2013. This text is a product of ASHRAE Technical Committee 5.3, Room Air Distribution, as assembled by the Underfloor Air Distribution subcommittee. The authors comprised industry experts on UFAD with regard to their specific fields. Thanks are due to the voting members of the committee, and to their employers, for the hundreds of hours volunteered. Special thanks to AirFixture, LLC, and Joseph Hullebusch for creating and coordinating many of the figures contained in this supplement.

The focus of this guide is to expand on issues related to operations, maintenance, and troubleshooting of underfloor air distribution systems that additional research and field experience has uncovered since the 2013 publication of the *UFAD Guide*.

James E Megerson, PE, LEED AP
Chair
November 2016

The following members contributed to this document

James E. Megerson, PE, LEED AP, Chair Design Mechanical, Inc.	Peter Strapp Tate Access Floors
William Reynolds Tate Access Floors	Jerry M. Sipes, PhD, PE Price Industries Inc.
Joseph Hullebusch Airfixture, LLC	Randy Zimmerman Titus
Bob Persechini RDK Engineers	Fred Bauman Center for the Built Environment (CBE)
Matt Bhumbla B.Eng, MBA, LEED® Green Associate Price Industries Inc.	Stefano Schiavon UC Berkeley
Kevin J. Gebke DuctSox Corporation	Kenneth J. Loudermilk, PE Titus

1 General Considerations

1.1 Introduction

Underfloor air distribution (UFAD) is an air distribution strategy for providing ventilation and space conditioning in buildings as part of the design of an HVAC system. UFAD systems use an underfloor supply plenum located between the structural concrete slab and a raised-floor system to supply conditioned air through floor diffusers directly into the occupied zone of the building. Thermal stratification, which is one of the featured characteristics of a UFAD system, allows using warmer supply air than that used with traditional overhead systems. The UFAD cooling load profile is different from a traditional overhead system because of the effect of the raised floor, particularly because systems using UFAD may typically have higher peak cooling loads than those using overhead air distribution. UFAD has several potential advantages over traditional overhead systems, including

- layout flexibility,
- enhanced individual thermal comfort control,
- improved ventilation efficiency,
- improved energy efficiency in suitable climates, and
- reduced life cycle costs.

UFAD is often used in office buildings, particularly highly reconfigurable and open plan offices, where raised floors are desirable for cable management. UFAD is appropriate for various different building types, including commercial office buildings, schools, museums, libraries, etc. UFAD systems have also been used successfully in performing arts facilities and other public meeting venues.

1.2 UFAD System Overview

UFAD systems rely on air-handling units (AHUs) to filter and condition air to appropriate supply conditions before it can be delivered to the occupied zone. Whereas overhead systems typically use ducts to distribute the air, UFAD systems use the underfloor plenum formed by a raised floor. The plenum generally extends 12 to 18 in. (0.3 to 0.5 m) above the structural concrete slab. Specially designed floor diffusers are used as the supply outlets. The most common UFAD configuration consists of a central AHU delivering air through a pressurized plenum and into the space through floor diffusers. The UFAD configuration may also incorporate fan-powered terminal units at the outlets, underfloor ducts, etc. Refer to the *UFAD Guide* (ASHRAE 2013) for additional information.

1.2.1 Life-Cycle Costs

A properly designed raised-floor system with UFAD can significantly reduce costs associated with frequent changes in office configurations, provided that program requirements do not change. This is a significant factor in life-cycle costs.

Every building has its own set of design constraints and application of UFAD will require evaluation to determine if they are applicable to a specific project. In addition to UFAD, life-cycle cost analysis must consider all components of the system, including electrical distribution (power, voice, and data). It is the consideration of both systems that leads to the ease of reconfiguration that affects life-cycle costs.

Life-cycle cost components include those that accrue over the lifetime of the building. Additional first costs can be offset by the reduction in ongoing costs over the life of the building. How many years of ongoing savings are required to offset additional first costs is left to the owner's judgment.

1.2.2 Energy Optimization and Performance

A performance assessment for a UFAD system should contain the following three primary elements:

1. ENERGY STAR® rating system. This relatively simple procedure uses one year's worth of monthly data for electricity, gas, and water consumption from utility meter records to provide a comparative assessment of the building's energy performance. These data are entered into the EPA Portfolio Manager calculation tool to obtain the ENERGY STAR rating (on a scale of 1 to 100) for suitable buildings. This process compares the building's energy performance with a large database of similar buildings nationwide (see www.energystar.gov/benchmark). If the building type under analysis is not included in the EPA list, it is still worthwhile to calculate the building's energy use intensity with EPA Portfolio Manager.
2. Temperature measurements. It is difficult to fully evaluate UFAD performance without measuring stratification and diffuser discharge temperatures. During a three- to four-day site visit, using a mobile measurement system to collect this data can result in a wealth of insights in how to proceed with adjustments (Webster et al. 2011).
3. Occupant satisfaction survey. This survey, administered via the Internet over a two- to three-week period, provides an assessment by the occupants of the quality of their indoor environment. The survey addresses all major categories of indoor environmental quality, including thermal

comfort, air quality, lighting, acoustics, cleanliness and maintenance, office layout, and office furnishings, as well as satisfaction with floor diffusers. The survey results can be compared to a large benchmark database maintained to assess relative performance (Frontczak et al. 2012).

ASHRAE's *Performance Measurement Protocols for Commercial Buildings* embodies all of these features (and more) and is a good way to evaluate performance (ASHRAE/USGBC/CIBSE 2010).

1.3 UFAD Air Distribution and Stratification

Thermal stratification is the result of processes that layer the internal air in accordance with relative density. The resulting air stratification is a vertical gradient with high-density, cooler air below and low-density, warmer air above. UFAD systems capitalize on the natural stratification that occurs when warm air rises because of thermal buoyancy. In a UFAD design, conditioned air stays in the lower, occupied part of the room, while heat sources such as occupants and equipment generate thermal plumes that carry warm air and heat-source-generated pollutants toward the ceiling where they are exhausted through the return air ducts.

The temperature stratification created by the UFAD system has implication for space set points. Most of an occupant's body is in an area that is slightly cooler than the temperature at the thermostat; therefore, current practice recommends raising thermostat set points compared to traditional overhead systems. The optimal ventilation strategy controls the supply outlets to limit the mixing of supply air with room air to just below the breathing height of the occupied space. The warmer and more polluted air transported by the thermal plumes is above the occupied space, out of the breathing zone. The air that the occupant breathes will have a lower concentration of contaminants compared to conventional, uniformly mixed systems.

Many factors, including the ceiling height, diffuser characteristics, number of diffusers, supply air temperature, and outlet airflow rate, affect the cooling load and ventilation efficiency of UFAD systems. Swirl-type and displacement-type diffusers have been shown to successfully create low air velocity in the occupied zone, whereas improperly applied linear diffusers created the highest velocity in the occupied zone, potentially disturbing thermal stratification and posing a potential draft risk. Additionally, many floor diffusers add an element of personal control within the reach of the occupant, because users can adjust the amount of air delivered by the diffuser. Refer to the *UFAD Guide* (ASHRAE 2013) for additional information on diffuser types and characteristics.

1.3.1 UFAD Design Tools for Zone Airflow Requirements

There are two available design tools for determining zone airflow rate requirements for UFAD systems. One was developed at Purdue University as part of ASHRAE research project RP-1522. The other was developed at the Center for the Built Environment (CBE) at the University of California Berkeley.

ASHRAE RP-1522 (Jiang and Chen 2012) developed a simplified tool available only to ASHRAE members that predicts the vertical temperature difference between the head and ankle of occupants, supply airflow rate for one plenum zone, number of diffusers, and air distribution effectiveness. The tool requires users to specify the zone cooling load and the fraction of the cooling load assigned to the underfloor plenum. It also allows users to input the supply air temperature either at the diffuser or at the location within the plenum where it enters the confines of the zone. The tool allows users to select from three types of diffusers.

The CBE tool predicts the effective cooling load for a UFAD system using the input of the design cooling load calculated for the same building with an overhead system. It also predicts the airflow rate, room temperature stratification, air distribution effectiveness, and plenum temperature gain for both interior and perimeter zones of a typical multistory office building using a UFAD system. The CBE tool allows the user to select from four different plenum configurations (series, reverse series, independent, and common) and five floor diffusers (swirl, square, displacement, and two linear bar grilles). An online version of the design tool is publicly available at the CBE website (www.cbe.berkeley.edu) (Schiavon et al. 2011).

1.3.2 Heat Transfer Pathways in UFAD Systems

Plenum supply air temperatures rise as the supply air travels through the underfloor supply plenum from the plenum inlet to the floor diffusers. This phenomenon is typically known as thermal decay. Plenum air temperature rise is caused by cool supply air coming into contact with the warmer concrete slab and raised floor. Elevated diffuser air temperatures lead to higher supply airflow rate and increased fan and chiller energy consumption. Typical temperature rise is 6.6°F (3.7°C), commonly ranging between 4.3°F and 8.5°F (2.4°C and 4.7°C) (Lee et al. 2012). This rarely happens on the ground floor with a slab on grade has less plenum temperature rise compared to the middle and top floors. Supply plenum air temperature rise thus has implications for the energy saving potential of UFAD systems and their ability to meet cooling requirements with supply temperatures above those of conventional overhead systems. Both energy and thermal performance can be improved in UFAD systems by ducting air to perimeter zones where loads tend to be the greatest. However, critics suggest that such underfloor ducting reduces the benefit of having a low-pressure plenum space, as well as adding design and installation complications when fitting ducts between floor tile pedestals. Decoupled sensible cooling devices such as radiant panels and chilled beams can also be effective in minimizing in both interior and perimeter area airflow rates.

1.4 UFAD and Energy

The energy assessment of UFAD systems is a not fully resolved issue, which has led to numerous research projects within the building science and mechanical engineering community. Proponents of UFAD point to the lower fan pressures required to deliver air in a building via the plenum versus via ducts. Typical plenum pressures are 0.0036 psi (0.1 inch of water column [25 Pa]) or less. The improvements in cooling-system efficiency inherent in operation at higher temperatures save energy, and relatively higher supply air temperatures allow longer periods of economizer operation. However, an economizer strategy is highly climate dependent and requires careful control of humidity to avoid condensation. Critics cite the shortage of rigorous research and testing to account for variations in climate, system design, thermal comfort, and air quality to question whether UFAD is able to deliver improved energy efficiency in practice. Limited simulation tools, the shortage of design standards, and relatively scarcity of exemplary projects compound these problems.

1.5 Applications

In addition to its frequent use in highly reconfigurable and open plan offices where raised floors are desirable for cable management, UFAD is also common in command centers, IT data centers, and server rooms that have large cooling loads from electronic equipment and requirements for routing power and data cables.

UFAD systems are not recommended for some specific facilities or spaces, such as small, nonresidential buildings, wet spaces (e.g., restrooms, pool areas), kitchens, dining areas, and gymnasiums, because UFAD may result in especially difficult or costly design. UFAD systems should also not be applied in laboratories because of its critical room-pressurization requirements and the potential migration of chemicals into the occupants' breathing zone as a result of their thermal plume formation.

Maintenance and Cleaning 2

2.1 General Guidelines

For raised-floor systems, it is important to consider the cleanliness of both the surface and interior of the system. Often, the system is both a floor as well as an air plenum for air distribution and cable management. Whereas most airborne contaminants and debris are heavy enough to remain at the bottom of the plenum because of the low air velocities within the plenum, it is important to check for contaminants that can be distributed throughout the floor plate. It is also important to look for contaminants whenever opening or accessing a raised floor. Panel surfaces should be properly cleaned quarterly, and the underfloor should be cleaned as needed based on regular inspections for the quantity of contaminants.

Take care to use proper cleaning techniques and solvents for the various specialty finishes that may be preapplied or field applied to the raised-floor tiles. Whenever applying a cleaning solution, always apply it first to a small area to see if there is any discoloration or staining before to applying it to the entire floor. Contact the original flooring manufacturer for care instructions before cleaning, and keep a record of the proper maintenance techniques.

Staff members who clean beneath the raised floor should be properly trained to be aware of and cautious around the power and data cabling as well as the HVAC equipment. Attempting to move these cables can cause an outage of data, power, or both.

2.2 Access Floor System Care and Maintenance

2.2.1 Protection from Overloading

With any raised-floor system, there is a limit to the amount of load per unit floor area that can be supported. Although many various metrics exist, it is imperative that the owner of the raised-floor system understand the upper limits in the existing system and avoid exceeding the safe allowable system loads. Additionally, the access floor must be protected from overloads and circumstances that exceed its specifications. Movement of heavy loads and equipment may require protection of the floor system through the use of plywood or other suitable load-distributing materials. Refer to the manufacturer's specifications to understand these limits.

2.2.2 How to Open a Raised Floor

Although there are various types of raised floors, the most popular method for opening a raised floor is by using a suction cup handle, shown in Figure 2.1. These can be purchased from many vendors, including local hardware stores, and are often available in those stores' sections for sheet glass or countertops. Other most commonly required tools are screw guns with a #3 Phillips head and torque-limiting function for removal of corner lock screws that hold panels in place on their pedestals.

It is recommended to not remove too many panels at one time and to be sure to replace panels as work progresses throughout the floor plate. Removing too many panels or not replacing panels can sometimes drop plenum pressure significantly, which can misalign pedestals and make reinstalling panels more difficult. Follow these general precautions:

- Take care to avoid inadvertently rotating pedestal heads and changing height settings when panels are removed. Although pedestal heads have antirotation mechanisms, it is still possible to bump heads out of alignment.
- Do not attempt to carry a panel by the lifter; the suction could break and allow the panel to fall from your hand.
- After removing cut panels installed at walls and around columns, reinstall them exactly where they came from. Interchanging panels that were precisely cut for perimeter locations can cause interior panels to be tight or loose in the floor and also cause the floor to be out of square.
- It is good practice to remove panels only where immediate access is required and reinstall them as your work progresses. When a number of panels must be removed, do not take out more than six adjacent panels at once OR remove only every other panel in a row.

Figure 2.1 Suction cup handle.
Courtesy Tate Access Floors, Inc.

- The first panel taken out must be pulled with a lifter. Adjacent panels can be removed with the lifter or by reaching underneath them and pushing upward. Kneeling on the floor while removing and replacing panels may avoid back strain.

2.2.3 Removing Floor Panels

In a corner lock system, back the screws out so they disengage the pedestal heads (the screws can remain in the panel to prevent misplacement). Place the lifter near an edge of the panel and push on the handle to form suction. Lift the edge of the panel (as if opening a door) until its edge is above the adjacent panel, then remove it completely by hand. Lay the panel on the floor and release the suction by squeezing the release bar directly above the suction cups. Do not slide panels across other panels, because this can damage tile finishes. Follow these special precautions:

- Do not use screwdrivers, pliers, or other objects to pry or lift panels.
- Do not disturb the level adjustment of the support pedestals and position of the stringers (lateral member used to connect the pedestals to each other) while floor panels are out of the floor system.
- When reinstalling panels with trim, use a carpet shoehorn to ensure the life of the trim.
- Always make a final check to see that panels are correctly in place, level, and secure.

2.2.4 Reinstalling Floor Panels

All panels should be reinstalled with the aid of a panel lifter to avoid injury. To reinstall a panel using a lifter, attach the lifter near an edge and lower that side into place. In a corner lock positioning system, maneuver the panel as necessary to align its corner lock holes with the holes in the heads. Fasten the corner lock screws so their heads are fully seated. Use no more than 30 to 35 in·lb (3.39 to 3.96 N·m) of torque. Do not attempt to reposition a panel on the understructure by dropping or kicking it into place.

2.2.5 Stringer Systems

A stringer grid can become racked (out of square) (Figure 2.2) when a group of panels is removed and the stringers are bumped, pulled, walked on, or incorrectly reinstalled when underfloor work is performed. These abuses will cause panels to be visibly out of square after they are put back into the grid. Correcting a large out-of-square condition must be performed by a professional installer. Be aware that 2 ft (0.6 m) stringer systems are more susceptible to racking than 4 ft (1.2 m) stringer systems. Anyone performing underfloor work should adhere to the following instructions to prevent stringer racking:

- Remove panels only where immediate access is needed, and reinstall them as your work progresses. The surrounding panels in the floor help to maintain grid squareness while underfloor work is performed.

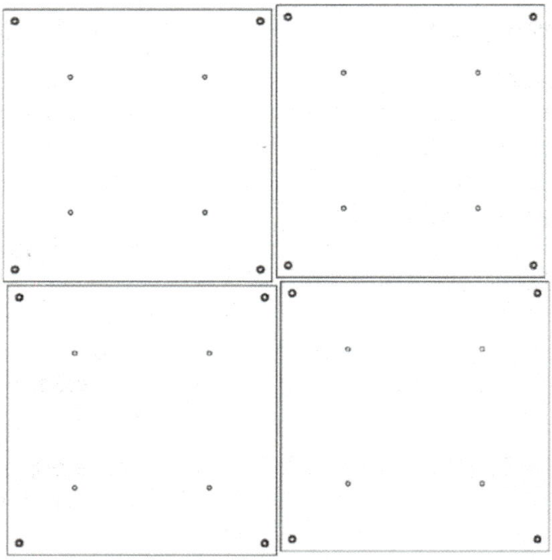

Figure 2.2 Plan view of racked access floor installation.
Courtesy Tate Access Floors, Inc.

- Keep foot and rolling-load traffic away from where panels and stringers have been removed. Nearby traffic can cause panel shifting and rotation of the stringer head connections.
- When removing stringers, remove only those stringers necessary for access, and replace them as your work progresses.
- When reinstalling a stringer, make sure the pedestal heads at each end are square to one another before final tightening of the screws.
- Never pull electrical cables against the pedestals and stringers.
- Do not walk on stringers or put loads on them when panels are removed from the floor.
- When removing cut panels at the floor perimeter and around columns, reinstall each panel section exactly where it came from. Perimeter panels are precisely cut for each location; interchanging them can cause floors to become out of square, and also cause interior panels to become too tight or too loose.
- Do not overtorque the stringer screws with your driver; use no more than 30 in·lb (3.39 N·m) of torque.
- Do not attempt to resquare the stringer grid by hitting stringers with a hammer or other objects; this can result in forming bumps on the stringer tops, causing panels to rock.

2.2.6 Adjustments to the Floor System

The floor owner or owner's representative should see a demonstration of the pedestal height adjustment and leveling mechanism by a representative of the floor provider before making adjustments to the system. Periodically inspect the installed floor and make adjustments.

2.2.7 Rocking Panels

Check to ensure that pedestal head gaskets are positioned correctly and/or stringers are securely fastened. Also ensure that no carpet pile or dirt prevents panels from seating correctly. Adjust the pedestal leveling nut after confirming that the pedestal is sitting flat on the subfloor. For persistent problems with tiled panels, try rotating the panels by 90 degrees.

2.2.8 Tipping at Perimeter Panels

Tipping generally occurs when the perimeter panel is improperly supported. Adjust the leveling nut on the pedestal support, raising or lowering the pedestal head until the panel sits firmly on the pedestal head or stringer.

2.2.9 Cleaning Floor Finishes

Always contact or refer to manufacturer-recommended methods for cleaning floors, because certain finishes may be susceptible to stain or discolorations. Because there is a nearly infinite number of different floor finishes, it is best to contact the manufacturer than to list them all.

Avoid using a broom or dry mop for cleaning, because these will only push dirt and debris around the floor and cause them to settle in another area or in the small seams between panels that sometimes form over time as panels settle. It is recommended to first attempt to clean any panel surfaces first by vacuuming to clean any large debris. Also be cautious of cleaning panels with very wet mops, because water may leak between tiles and damage or destroy data and power wiring or HVAC equipment beneath.

2.3 UFAD System Maintenance

UFAD systems generally require minimal system maintenance. This is generally covered by regular system cleaning, comparable to that needed for any air distribution system. This section outlines special considerations for maintaining a UFAD system.

2.3.1 Plenum Cleaning and Timeline

After the subfloor is cleaned of dirt and debris from construction, further accumulation will be minimal and can be removed with a shop vacuum. Because the cavity is enclosed, significant buildup may take several years to develop. Determine when cleaning is necessary by occasionally inspecting the subfloor. The required frequency of cleaning is estimated at two to three years based on the rate of buildup observed. HEPA-filtered vacuums are recommended for cleaning to prevent dust and dirt from being released into the room air.

Cable cutouts in floor panels should be sealed with grommets or trim-and-seal components to prevent entry of dust and dirt into the cavity.

If liquid enters the floor cavity by spillage or water pipe leakage, vacuum extraction is the most effective way of removing it. The drying process can be accelerated with dehumidifiers, fans, and/or heaters. Extraction and drying should occur as soon as possible.

For areas subject to excessive moisture or potential water leakage, water leak detection systems made for use in access floor cavities are available. These systems monitor the subfloor in places where water leakage may otherwise go undetected.

2.3.2 Maintaining System and Equipment Accessibility

The raised access floor allows for easy accessibility to UFAD equipment and components for maintenance and serviceability. It is important to maintain this accessibility when any work is performed under the floor and avoid adding barriers to access equipment. This includes installing ceramic tile, underlayment, or other floor finishes that prevent access to equipment below the raised-access floor.

2.3.3 Cleaning Air Diffusers and Terminals

Underfloor air plenums contain varying levels of equipment, depending on the type of plenum and spaces serviced. Pressurized plenums contain the least equipment. Special areas, such as perimeter zones, conference rooms, and zero-pressure plenums contain more equipment, such as fan-powered boxes. As with any equipment location, adequate clearance must be provided for maintenance, repair, and removal. The operation and maintenance (O&M) organization should review the equipment layout to ensure clearance is provided.

Dirt and debris that fall into floor air diffusers will be collected by the baskets beneath the diffuser cover. Diffusers, electronic air terminals, and power distribution boxes may be cleaned whenever excessive dirt or debris develops in the collection basket or steel housing. The air-handling system should be turned off during maintenance of electronic air terminals to avoid possible injury. Loose debris should be vacuumed out of the units. Some equipment may be supplied with filters; with underfloor air plenums, these are often unnecessary because of pretreated supply air pressurized plenums. Refer to equipment specifications for additional requirements regarding regular maintenance requirements. Grilles and air diffuser baskets may be removed and cleaned using a soft cloth soaked with warm water and a mild cleaner, and dried with a similar cloth.

2.3.4 Other Systems and Equipment

UFAD plenums are almost always used to provide raised access floor benefits. For this reason, O&M access to other than UFAD systems and equipment, such as electrical and cabling, must be ensured as well.

UFAD Control and Optimization 3

3.1 Controllability of Systems

The control and optimization of thermal comfort in the occupied zone and the amount of thermal stratification during cooling operation is crucial to system design, sizing, energy-efficient operation, and comfort performance of UFAD systems. This section presents recommended control strategies for effective system operation specific to UFAD systems.

Depending on the space requirements, the controllability of set points and allowable ranges may differ from zone to zone. In some, the control of parameters and values is primarily for occupant comfort. In others, more restrictive values are required to protect property and processes. This chapter focuses on control for occupant comfort and proper pressurization of the building during both occupied and unoccupied conditions.

3.2 Thermal Mass

It is important to consider the thermal mass of the supply air plenum when changing set points. The materials of the supply air plenum will absorb energy over time. It is recommended that changes to supply air temperature be made in increments of 1°F to 2°F (0.5°C to 1.0°C) at a time. Results from changes may take up to several hours or days, so sufficient time should be allowed to obtain the results desired. Making changes larger than this may cause overshooting of results and create other occupant comfort issues.

3.3 Trend Analysis

Once the UFAD system is commissioned, it should perform consistently unless altered or subjected to physical degradation. Current building automation systems have the capability to trend operational data and optionally use fault detection and diagnostic analysis. These tools should be used by the operators to proactively identify or anticipate impending problems and failures.

3.4 Common UFAD Control Principles

Because of the number of different UFAD system configurations, this section does not cover all possible control scenarios, but is intended to introduce some of the basic control strategies that are frequently used in UFAD systems.

3.4.1 Pressure Reset of Plenum Supply Air

As stated previously, the majority of supply air diffusers used for UFAD applications are pressure dependent (refer to Figure 3.1 for a typical static pressure versus air quantity diffuser curve). In other words, given a specific static pressure, the diffuser will deliver a specific quantity of air. Depending on the system design, control of air volume may be adjusted to match the space load at any given time by controlling the plenum pressure with respect to the space. This may be important should the load be intermittent or variable. Modulating fan speed or discharge air dampers to maintain a predetermined supply air static pressure is a common control strategy for this scenario, especially during off-peak load conditions.

Typically, this is done by resetting the supply air plenum pressure based on return air or space temperature (see Figure 3.2 for an example of supply air pressure reset schedule). The return air temperature or space temperature is used as an indication of the current load in the space. As the temperature rises, it indicates that there is more heat load in the space, requiring more supply air to satisfy the load. This can be especially important if manual diffusers are used in the space or if there is more leakage through closed diffusers or other openings into that space than anticipated. Therefore, the cooling load present is much lower or higher than the current capacity being delivered.

Note that careful evaluation is needed to determine the best applicable strategy for controlling air volume to the space, especially if a common plenum is used to supply air to areas with different load profiles or schedules. For example, if the supply air pressure is reset down because of overcooling of an open office floor plan, occupied conference rooms or office spaces served by the same plenum may not be able to maintain space temperature because of the lower airflow rates. In this case, it may be necessary to provide more diffusers in these zones to maintain adequate cooling with lower-than-optimum design pressures. It is also possible to use zone booster fans, plenum dividers, or other means to maintain

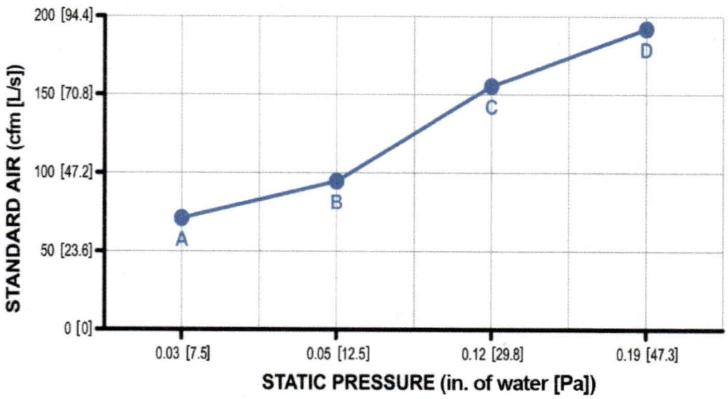

	cfm	L/s
A	66.8	31.5
B	97.0	45.8
C	154.8	73.1
D	190.1	89.7

Figure 3.1 Typical pressure-dependent diffuser operation.

the proper volume of air to these spaces when system pressure has been reset down.

Diffuser selection should take into account different plenum pressures. For example, it would not be recommended that diffusers of different operating pressures (e.g., type "A" full flow at 0.03 in. of water [7.5 Pa] and type "B" full flow at 0.05 in. of water [12.5 Pa]) be used with the same plenum without careful evaluation.

3.4.2 Temperature Reset of Supply Air

In some designs or climate zones, it may be permissible to use a supply air temperature reset schedule based on the load profile of the space (see Figure 3.3).

Much like supply pressure reset, the air temperature can be changed while the air volume (i.e., plenum pressure) remains constant, to better match the current load for a given zone as indicated by return air or space temperature. This is typically done by resetting the leaving-air coil temperature or modulating return air bypass dampers or reheat coils.

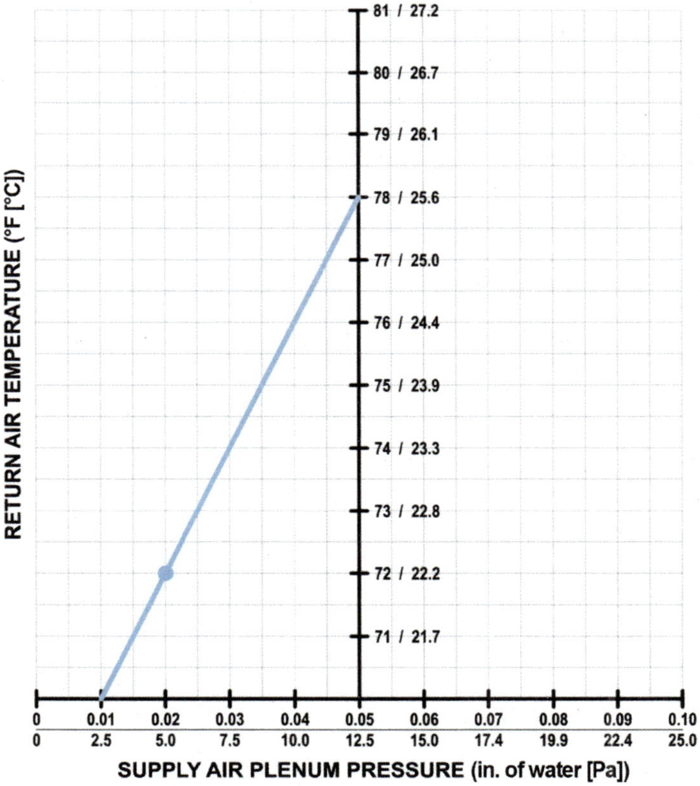

Figure 3.2 Typical supply air pressure reset schedule.

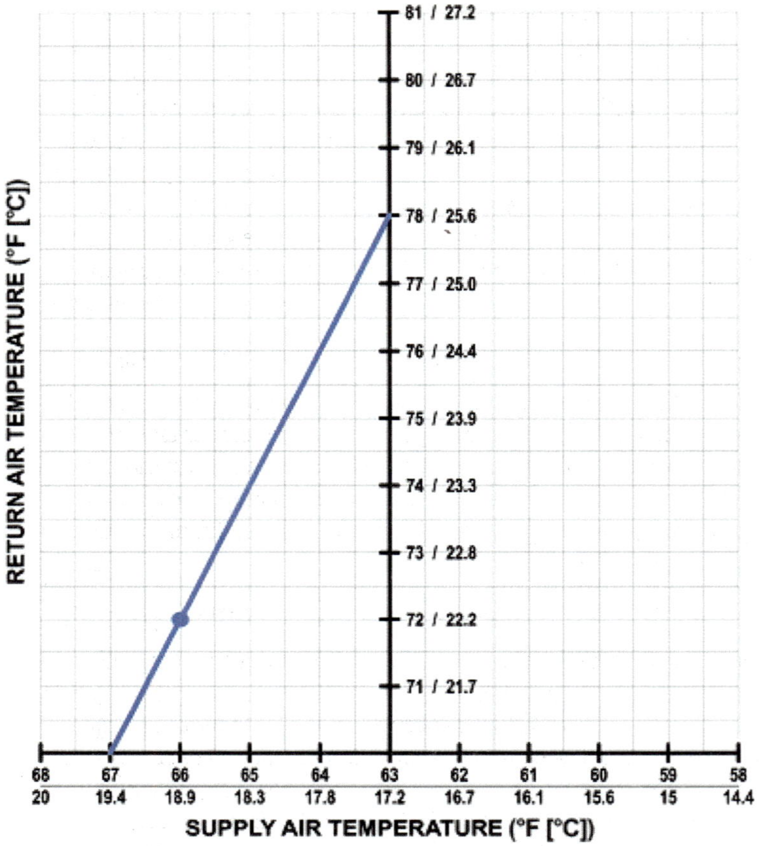

Figure 3.3 Typical supply air temperature reset schedule.

Temperature reset sometimes works better when minor adjustments are made on a seasonal basis (e.g., a supply air temperature of 65°F [18.3°C] when outdoor air temperatures are above 50°F [10°C], and a supply air temperature of 68°F [20°C] when outdoor air temperatures are below 50°F [10°C]).

The same caution as for pressure reset should be considered. It may be possible to satisfy the load for one space with temperature reset but negatively affect another space served by the same plenum. It may be necessary to use plenum dividers, diffuser counts, or booster fans for these situations.

In addition, if used aggressively, temperature reset could use more energy and make space temperature swings more dramatic. Note that energy will be absorbed and released by the plenum materials (e.g., concrete slab), which could delay or even exacerbate the space conditions.

3.4.3 Combined Temperature/Pressure Reset

This strategy may be used concurrently for faster zone reaction to load variation. However, the same concerns as expressed in Section 3.4.1 should be eval-

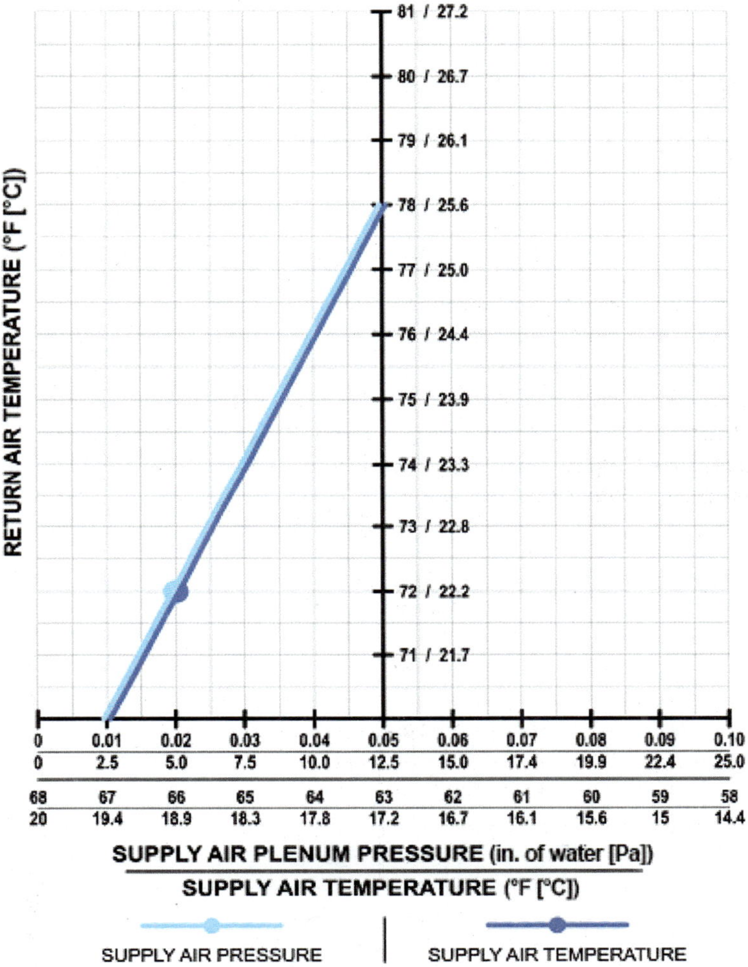

Figure 3.4 Combined supply air temperature and pressure reset schedule.

uated. It is also recommended that the reset schedules coincide with one another so that changes made to the pressure and temperature do not negatively affect the other. The potential for this combined strategy is continual hunting of dampers and valves to satisfy both scenarios. Figure 3.4 is an example of a temperature and pressure reset schedule that coincides with return air or space temperature feedback.

3.4.4 Variable-Frequency Drives (VFDs) and Other Variable-Flow Control

When using a variable-volume UFAD air distribution strategy, it is important to provide a means for modulating the amount of supply air to the plenum for energy savings. This can be done by using VFDs or variable-flow-control dampers, modulating based on the supply air plenum pressure.

If multiple floors or areas are served by the same air handler, both the VFD and floor dampers should operate in series with one another. It is recommended that the air handler maintain the minimum duct static necessary to supply all plenums with maximum airflow at the building's peak demand (taking diversity into account). Individual supply air plenum (zones) dampers modulate to maintain the supply air pressure needed for the load conditions experienced by each zone. The fan speed of the air handler should be modulated with a VFD as the pressure in the main duct fluctuates due to zone dampers opening and closing to maintain pressure in the supply air plenum (see Figure 3.5). It is also recommended that the proportional-integral-derivative loops used are such to prevent dampers and VFDs from hunting.

It is generally recommended to serve a large plenum with multiple air injection points. This is done to minimize the effect of potential plenum supply

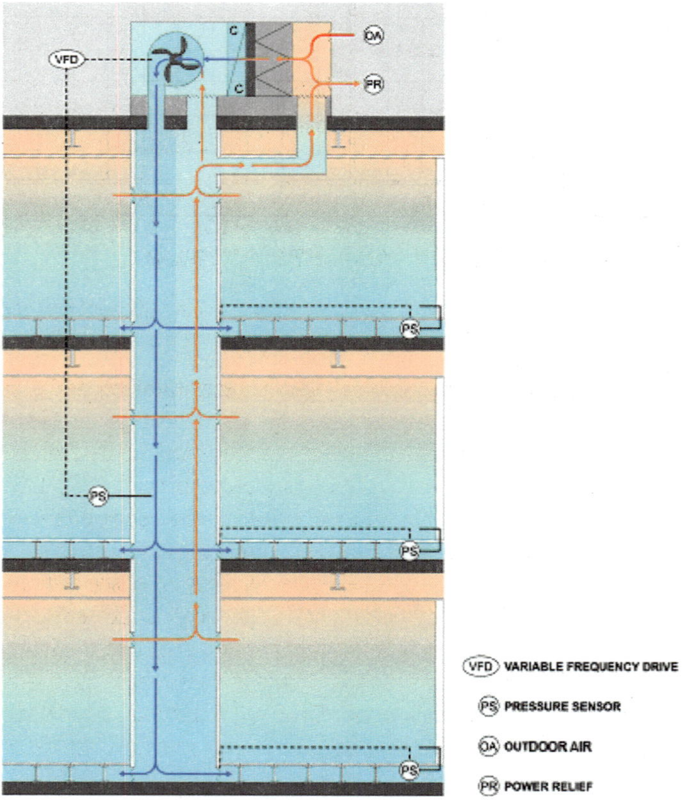

Figure 3.5 System schematic: combined VFD/zone control damper system.
Courtesy Larson Binkley, Inc.

air temperature rise. Each of these injection points should use a control damper controlled by the average pressure in the plenum. It is recommended that all dampers serving the same plenum be synchronized and modulated together to one common average pressure.

3.4.5 Economizer Cycle

The supply air temperature for UFAD is higher than for overhead systems. Therefore, an economizer cycle can be used over a wider range with overhead systems, resulting in energy savings. However, an economizer's effectiveness depends on a building's location. Each building must be evaluated to determine if economizer operation leads to savings in its local climate.

One hundred percent free cooling (i.e., no cooling coil output is required to maintain supply air temperature, not to be confused with 100% outdoor air damper position) only happens when outdoor air temperature is less than or equal to supply air temperature (which can be reset upward in mild weather). The typical UFAD supply air temperature range is 63°F to 66°F (17°C to 19°C). It is reasonable to use 100% outdoor air (OA) for cooling the building when it is at or below the design supply air temperature set point. Note that, to use OA for free cooling, the OA must have a low enough enthalpy.

3.5 Acoustics

Because of the elimination or minimal use of ductwork in underfloor plenums, noise generated from the operation of a UFAD system can be substantially less than that from a conventional ducted overhead system, which can lead to quieter spaces but also speech privacy issues. This reduction in commonly found levels of background HVAC noise may create a situation where active sound masking or other acoustic design measures may be required. The small volumetric fans used for fan-powered diffusers typically have lower noise levels than larger fan terminal units. Attention should be paid to fan-powered devices in terms of manufacturers' specifications, correct installation, and proximity to surrounding diffusers, and other acoustic measures that can be taken.

Raised-floor panels usually comprise a metal casing (typically galvanized steel) and core material or a metal-framed concrete panel. By introducing a solid core, such as lightweight concrete or high-density particleboard, problems with hollow-sounding floors are avoided and acoustic insulation performance is improved. In terms of reaching a compromise between a soundtight installation and the flexibility of frequent and easy removal of floor panels, any seals/ mechanical locking configurations must be durable enough to avoid degeneration from wear and tear. In addition to the panel itself, carpet-finished systems offer a good level of acoustic insulation by damping vibrations.

Sound transmission between adjacent rooms with floor diffusers served by the same plenum is sometimes cited as a concern. Careful placement of diffusers should reduce noise transmission. Avoid placing diffusers directly adjacent to each other on opposite sides of a room partition. Diffuser locations should also be chosen to avoid direct air paths from discharge ducts under the floor.

3.5.1 Discharge Openings in the Supply Air Plenum

Air is usually introduced into the supply plenum through a duct branch from a supply air shaft. It may also be done by an air handler that discharges directly into the supply air plenum. Take care to ensure that the entering air velocity does not produce objectionable noise for space occupants. Air velocities entering the plenum should not exceed 3000 fpm (15 m/s). At this velocity, sound levels have been shown to be maintained at noise rating level NC 35 or below. The following are a few advantages of maintaining air velocities at approximately 3000 fpm (15 m/s):

- Smaller ductwork is desirable because of limitations created by the raised floor. Raised floors are typically supported on pedestals that are 24 in. (600 mm) on center, which means ductwork can be no wider than approximately 22 in. (559 mm).
- First costs can be reduced because of smaller and fewer ductwork branches below the raised floor.
- The effects of plenum temperature rise are decreased because higher-velocity air flows to the exterior of the building faster, minimizing heat gain from the plenum itself.
- Supply air duct lengths can be minimized, because the air will move faster for a longer distance under the floor.
- Duct static pressure increases produced by the higher velocities are offset by shorter runs of ductwork.

At higher UFAD supply air entering velocities, the temperature gain of the supply air plenum is less from the injection point to the perimeter. Therefore, supply air inlet velocity of 3000 fpm (15 m/s) is acceptable for UFAD applications.

3.6 Plenum Integrity

For the plenum to act as an air delivery means throughout the life of the building, its ability to contain the air must be maintained.

3.6.1 Leakage Testing

Opinions differ in the UFAD community on whether supply air plenum leak testing is absolutely necessary. This is mainly because of the cost of testing and the amount of time needed to properly perform the test. Some believe that leak testing can be avoided if diligent oversight of the plenum construction is performed to ensure that the plenum is installed properly. Another approach is to forgo leak testing if the system can make appropriate pressures as intended. This section focuses on the process of leakage testing rather than whether it is necessary.

If the initial plenum installation is leak tested during the testing, adjusting, and balancing and commissioning processes, this value can be used as a baseline to compare future leakage test results. A leakage test can be performed both before and after major reconfigurations. The pretest provides an indication of the

Figure 3.6 Example signage.

degree of increased leakage over time and helps determine how much effort must be dedicated to inspecting and resealing the plenum. Note that, as the space is used, leakage can increase because of many factors, including the following:

- Degradation and shrinkage of sealing materials
- Settling and movement of building materials, resulting in cracks and gap widening
- Improper replacement of floor tiles and floor utility assemblies
- Breeching of the plenum

The last factor, plenum breeching, is probably the most important. The O&M personnel must fully understand the importance of maintaining plenum integrity when the space is modified or utilities are added or rerouted. It is recommended that a permit system be developed for all plenum-related work and that all anticipated plenum penetrations be specifically identified and inspected by the building owner's representative. In addition, signage, such as that shown in Figure 3.6, should be posted in the vicinity of existing penetrations that are likely to be disturbed in the future.

3.6.2 Isolation of Thermostat from Pressurized Plenum

Typically, space thermostats are mounted on a wall that sits on the raised floor or on the slab below. It is important to isolate the thermostat from direct contact with the supply air from the plenum. If using a conduit and back box for the thermostat, the conduit end where the wire exits should be sealed airtight. It is also important to make sure that the wall cavity is also sealed and isolated from air entering the cavity. If not properly isolated, it is possible to get an inaccurate thermostat reading, causing space comfort problems (refer to Figure 3.7).

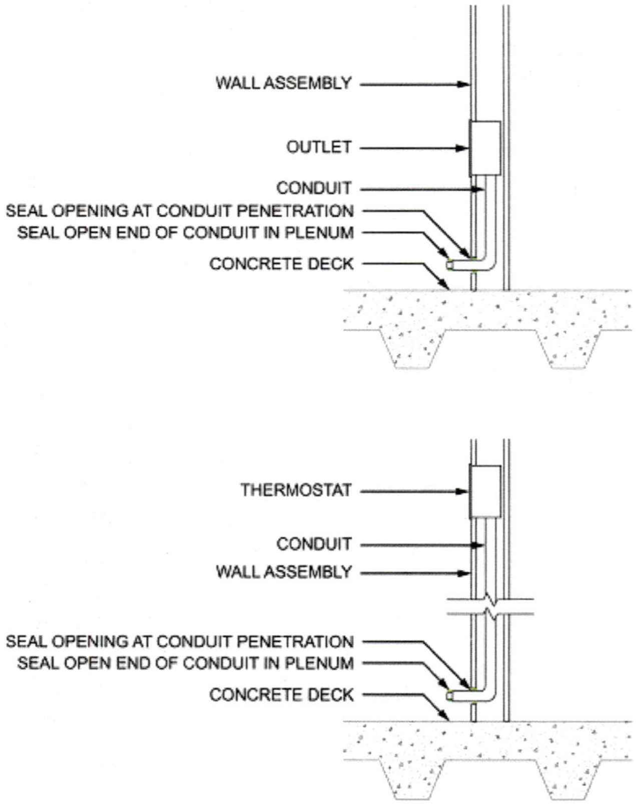

Figure 3.7 Construction details for open ends of conduit in plenum spaces.
Courtesy Tate Access Floors, Inc.

3.7 Occupant Satisfaction

Conducting a periodic indoor environmental quality occupant satisfaction survey could be key in assisting with troubleshooting operational problems with the UFAD system.

3.8 Occupant Training

It is imperative that the occupants not only understand how to control temperature and airflow but also grasp the basic principles of UFAD. Specifically, they must understand that room response is more gradual than with other systems.

Refer to Section 4.1.2, Diffusers, for additional details related to diffuser location and adjustment.

System Moves, Adds, and Changes 4

The importance of careful design of UFAD systems cannot be overemphasized. Once a system is installed, there are few available adjustments that can dramatically alter the energy and comfort performance to correct for design flaws. Other types of remediation (e.g., replacing/adding diffusers) can be expensive. Even commissioning is complicated by the high degree of interaction between various elements of the system and the need to achieve balance between energy and comfort concerns. Among obstacles to commissioning are the following:

- Preoccupancy testing is impractical; without normal loads in the space, stratification cannot be created.
- Midseason testing is difficult because of changeovers between heating and cooling in the same zone or between zones. Cooling the interior while heating the perimeter complicates testing because of interactions between the two temperatures.
- Load conditions are difficult to measure.
- Elaborate testing procedures are impractical because of their complexity, the inability to replicate test conditions in the field so that comparisons can be made to previous tests, and the high cost of testing and disruption to tenants.
- Unresolved problems with HVAC components and the building management system (BMS) controls (i.e., lack of complete functional testing and commissioning of BMS).
- Sequences of operation that do not conform to specifications, are inoperable, or are inappropriate for UFAD.
- Adjusting supply plenum distribution is often impractical because of difficulties in access and inability to predict the outcome.
- Wide arrays of approaches are used to implement a UFAD system at the tenant level, AHU level, and at plenum airflow distribution configurations. These affect interactions between interior and perimeter zones. Although there appears to be some degree of decoupling between room performance (at least on a large area basis) and the plenum temperature distribution, experience has shown that balancing comfort and control between these zones because of the high degree of interaction between them represents one of the major challenges for tuning these systems.

Assuming commissioning is done well, energy and comfort performance can be affected mostly by the following parameters: AHU supply temperature, room set points, diffuser placement compared to occupant and adjustment, and adjustment (lowering) of minimum volume settings. The goal is to improve comfort and increase stratification to optimize energy performance. These adjustments indirectly affect stratification, and each has side effects that could present other difficulties or make energy or comfort performance worse. The most effective way to ensure optimum performance is to design the system with adequate number and type of diffusers; this appears to ensure more uniform dynamic performance at relatively high stratification. For systems with serious design flaws, adding, adjusting, or relocating diffusers may be the best way to optimize performance and reduce comfort complaints. Some situations may require different approaches than others. For example, if terminal equipment is undersized (a common problem when plenum supply air temperature rise is not properly factored into the design calculations), lowering AHU supply temperature should be the focus (although increasing room set point will also help). However, this can exacerbate problems of comfort in interior zones (another common problem in UFAD systems).

Research results (Schiavon et al. 2015; Hoyt et al. 2015; Schiavon and Melikov 2008) indicate that increasing the room set point results in increased stratification and lower energy use, but the savings depend strongly on climate. Lower energy use is not unexpected, because increasing set points in any system lowers airflow requirements. It has been shown that fan energy savings contribute to cooling energy savings only in mild climates where there are a large number of economizer hours. On the other hand, UFAD systems may be better suited for using an increased set point strategy, because stratification creates lower effective temperatures in the occupied zone than overhead systems for the same thermostat setting. Also, UFAD provides some degree of personal control that can mitigate comfort problems that may occur at higher temperatures.

Adjusting relocated diffusers may be an underused strategy to mitigate comfort problems. Survey studies (ASHRAE 2003) have shown that few people adjust (sometimes they do not know how) their diffusers and often they are unaware that they can be moved easily. Occupant training is essential to optimizing the performance of these systems. This applies to operation personnel as well.

Lowering minimum volumes can be an effective way to mitigate comfort problems in interior zones. However, there is concern that this approach could negatively affect ventilation performance. Each system must be evaluated carefully beforehand.

4.1 System Adds or Moves

System adds or moves can occur frequently because of the modular nature of the raised-floor system. These changes typically can be completed by on-site facility engineering staff and do not require assistance from outside contractors.

Proper safety procedures for the specific equipment being moved should always be reviewed before starting any work.

4.1.1 Service Distribution Boxes

Service distribution boxes are typically designed with plug-and-play connectors to allow simple relocation of service boxes and wiring. If the specific device includes "first break/last make" grounding plugs, moves and adds can be completed under live power. If this feature is not clearly defined, electrical changes may need to be completed by an electrician.

4.1.2 Diffusers

With computer-aided design, interior layouts can be accurately planned, particularly the locations of diffusers in the workstations. However, because of installation tolerances and adjustments, workstation encumbrances, and personal preferences, some postinstallation diffuser relocation is typically required.

The first step in ensuring occupant satisfaction is to optimally locate the diffusers. A typical 8 × 8 ft (2.5 × 2.5 m) workstation probably has eight 24 in. (600 mm) tiles that are available for diffusers. Using an offset tile diffuser (e.g., located in one 12 × 12 in. [300 × 300 mm] quadrant of the tile) allows a significant amount of outlet adjustment. It is very important that the installer work with the O&M team and the occupants to ensure that initial diffuser location is optimized. In cases of extremely sensitive occupants, the O&M team may want to consider a minipilot with the occupant. This can be accomplished by leaving some carpet tiles out and taping the floor tile joints to facilitate moving the diffuser location until the occupant is satisfied.

Generally, two types of diffuser categories are available for installation and use in the raised-floor environment: manual and automatic variable-air-volume (VAV) diffusers. Manual diffusers are typically used and located in areas where a constant air volume with minimal daily adjustment is needed (e.g., hallways, open areas, café areas) and next to occupants' desks, where the user can be provided with the feature of personal control of their environment. Automatic VAV diffusers are typically used in locations with a high rate of turnover of cooling or heating requirements (e.g., conference rooms, perimeter areas, lobby areas).

4.1.3 Access Floor Panels

Raised floors are available in variety of material types, load ratings, and finish options, but are generally approximately 24 × 24 in. (600 × 600 mm) in dimension. These tiles are designed to be interchangeable within a specific area or within an entire building. This allows relocating equipment and raised-floor panels throughout the office space.

4.2 System Changes

Often, a change in system design may be required after the building is use and during occupancy. With a UFAD system, the cost and impact of completing these changes is mitigated because of the ability to easily move or add equipment. However, many changes or considerations are required to avoid affecting the performance of the entire system. Many of these considerations are identical to those for overhead systems, with a few exceptions, as discussed in the following sections.

4.2.1 Occupancy Type Changes

Occupancy type changes include any change where one or more areas of an office are converted from one type to another. In situations where the heat load and anticipated occupancy have changed because of this, it must be considered how the change will affect the design of the UFAD system. Areas which previously only required manual diffusers, such as open office areas and small individual offices, may now be configured to serve as conference or meeting rooms requiring VAV diffusers to accurately control the amount of cooling or heating provided based on temperature, CO_2, or occupancy.

4.2.2 Zoning Changes

Zoning changes typically consist of a change in controls of air-handling equipment to better serve the occupants of the spaces (e.g., changing a thermostat location or a zone definition to better suit the occupancy of the current office configuration). It can also consist of the addition of a manual diffuser into a floor plan where an occupant may need additional control of the air supplied in their specific area. Lastly, training of occupants on proper use of thermostats and diffusers is essential for peak system performance and occupant satisfaction. Occupants are often uneducated regarding the ability to adjust personal diffusers or local thermostats to meet comfort demands.

4.2.3 System Capacity Changes

System capacity changes can potentially have broad impact on the capacity demanded of air-handling equipment. When significant changes in demand occur as a result of increased or decreased occupants or increased spaces, it may be necessary to reevaluate the design capacity of the HVAC system supplying the UFAD plenum. Some indications that these considerations may not have been properly considered include an increase in hot/cold call complaints or difficulty maintaining a controlled temperature in a conference room or perimeter zone.

If a supplemental system has been introduced to a space being served by UFAD, the returns back to the plenum must be evaluated so that the static pressure in the room is neutral to allow the air to discharge out of the floor diffusers.

Any cabling, conduits, and ductwork added under the raised floor must be installed so as not to impede air movement.

Troubleshooting Common Issues 5

Operation of the UFAD system should be periodically monitored. Early detection of abnormal operation is important to preventing poor thermal comfort and other operational deficiencies, such as excessive energy use.

5.1 Red Flags

It is often possible to identify problems with occupant comfort before complaints are made, by observing a few common occupant behaviors, as discussed in the following sections.

5.1.1 Book/Box on a Diffuser

It is common for occupants sitting inside the clear zone of a floor diffuser to cover it with a book, tape, paper, etc. to prevent airflow from the diffuser. This indicates a comfort problem resulting from diffuser location, the location of the occupant, or operation of occupant control. In some cases, the diffuser may not allow shutoff from the supply air plenum, or the occupant may not have sufficient training to know how to shut off the diffuser.

5.1.2 Diffusers under Desks

It is not generally advisable to locate diffusers directly under an occupant's desk or chair. Most diffusers have a prescribed clear zone that manufacturers recommend so that air velocity from the diffuser is not objectionable to occupants.

5.1.3 Condensation on Windows

Although not necessarily a problem associated with UFAD systems, it is possible that condensation can accumulate on exterior windows. This can be caused by problems not directly associated with UFAD systems, such as poorly insulated glazing or window frames that leak and/or lack thermal breaks. Generally this occurs when the window or frame reaches a temperature at or below the dew-point temperature in the space. The first means of troubleshooting this problem is to take a humidity reading in the space as well as the surface that is condensing.

5.1.4 Furniture-Only Moves

It is very common for furniture moves to be made in the space without consideration to diffuser locations. Furniture should not be moved to where it is located on top of a diffuser. Other considerations are to move the diffusers so that they are not under furniture or within the clear zone for an occupant.

5.2 Cold Interior Zones

A common occupant complaint in a poorly designed system is cooler interior temperatures. Several issues can cause this situation.

Air entering the space in an uncontrolled way ("Category II leakage") can cause overcooling of the space before occupancy. It is paramount to find where this is occurring and make corrections. In some cases, it can be from manual diffusers left open when the space is unoccupied. One way to correct for this is to use either a supply plenum pressure or temperature reset strategy (see Chapter 3, UFAD Control and Optimization) and optimize the system start time for each day.

Another possible cause is locating too many return diffusers in the interior zone, causing short circuiting of the supply air closer to the central supply air shafts in the space. A relatively easy solution is to relocate interior return grilles to the perimeter spaces if a common return air plenum is being used.

5.3 Inadequate Perimeter Cooling

If temperatures at the perimeter of the building are warmer than desired, possible causes are inadequate airflow or too high a supply air temperature.

5.3.1 Inadequate Airflow

A load analysis may be needed to determine if the number and size of diffusers are adequate to deliver the correct airflow for cooling. If not, it is relatively easy to move, add, or change diffusers in the space to get proper airflow.

5.3.2 Air Too Warm for Cooling

This is generally easy to determine by taking temperature readings of the air exiting the diffusers or where air enters the supply plenum. Once it is determined that supply air temperature is higher than desired, it is possible to determine a cause and solution.

5.3.3 Determining Causes

Determining the cause of elevated temperatures at the perimeter can be a challenge. The following are a few items to consider while making an assessment:

- How is the return air configured in the space? It is essential to determine if the number and size of return air grilles and openings above the space are adequate to allow sufficient return air at a static pressure drop less than that of the supply air plenum pressure. If air cannot get out of the space, it will not supply airflow to the space at the correct quantity.

- It may also be an advantage to have the return air grilles located at the perimeter rather than the interior wall of the space. Allowing the heat plume from the exterior wall and glass to rise from the space and directly into the return will prevent it from being mixed back into the space by the supply air diffusers. Refer to Chapter 11 in the *UFAD Guide* for additional information.
- How is return air configured on the floor below? If return grilles are located in the interior of the space below, it's possible that a dead zone could be created in the return air plenum that promotes temperature gain in the return that will eventually transfer through the supply air slab causing temperature gain in the supply air on the floor above. Moving the return air grilles to the perimeter on the floor below may help alleviate this problem. Refer to Chapter 11 in the UFAD Guide for additional information.
- If the distance from where air enters the supply plenum to where it exits the farthest floor diffuser is greater than 50 ft (15 m), it may be necessary to install a duct extension below the floor or increase discharge air velocities to get the colder air closer to the perimeter. Refer to Chapter 11 in the *UFAD Guide* (ASHRAE 2013) for additional information.
- Solar radiation through perimeter glazing can be a significant cause of occupant discomfort. This is very difficult to remedy, because the solution may not be desired by the occupant. Installation of interior or exterior shading should alleviate this problem, even if the shades or blinds are not fully closed. Other solutions may include window tinting or other treatments to reduce the amount of radiation entering the space. This can be very effective in combination with locating the return air grilles at the perimeter.

5.4 Over- or Understratification

For some diffuser types, understratification and a lower space temperature is a sign that too much air is being delivered to the space. It may also be possible that the diffusers are causing air to mix at a higher elevation in the space than needed. Generally, decreasing the airflow to the space will correct this issue.

Overstratification in the space or if the temperature is cooler towards the floor but the space is not meeting the supply air temperature could be a sign that the space is not getting enough airflow. It may also be possible that there is not enough return air from the space.

5.5 Noise

It is not uncommon to hear concerns that buildings with UFAD are too quiet. This is generally because of the low pressures and velocities in the supply air plenum. In contrast, a typical overhead ducted system has higher velocities and pressures, causing background noise that can mask other ambient noise in the space. It is possible to address this for UFAD by using a sound masking system, whether throughout the building or only in problem areas.

In most cases where noise in a UFAD system is objectionable, it is the cause of a mechanical failure or misapplication of a fan-powered unit located below the floor. Every situation should be evaluated on a case-by-case basis.

5.6 Leakage

There are several indications that the supply air plenum is leaking. The most obvious is the inability to make supply air plenum (SAP) pressure. The more difficult determination is whether it is Category I or II leakage (refer to Chapter 3, of the *UFAD Design Guide* [ASHRAE 2013]). Performing leakage testing as discussed earlier can help identify the type of leakage. In general, here are several key symptoms of leakage:

- Inability to hold SAP pressure. If this is a new occurrence, it may be an isolated event. Look for floor tiles removed in data closets or electrical rooms. If recent work has been done involving modifications within the SAP, look for leaks that could have been caused as part of that work.
- Cold walls. This is an indication that air is leaking into the void space in a wall.
- Thermostat reading lower than actual space temp. Air is leaking into a wall cavity near the thermostat, or air is leaking up the conduit holding the thermostat wire. Sometimes the air can be felt escaping around the thermostat.
- Return air colder than average space temperatures. This indicates that cold air is bypassing into the return air plenum. Look for passages through walls or column wraps where this can occur.
- Space overcooling. This can indicate Category II leakage. Look for leakage through electrical boxes, closed floor diffusers, between raised-floor tiles, around wall moldings, etc.

Documentation 6

Although access, maintenance, and repair in UFAD plenums is generally easier than in overhead systems, this is predicated on the maintenance worker having ready access to the plenum. UFAD plenums are not conducive to exploratory access; that is, a worker cannot simply lift a ceiling panel and look around, as in an overhead system. Therefore, accurate and well-maintained as-built drawings are required that specifically identify the following:

- Equipment locations
- Equipment access panel and/or maintenance component locations, to help determine the approach floor tile(s) to be removed
- All shutoff and isolation valve locations

If accurate locations for systems and equipment are known, maintenance personnel can access these items and cause less disruption of operations than they would with overhead systems.

All manufacturer information must be maintained by the building O&M team and made available to any outside contractors that service the floor or the plenum.

6.1 Equipment Locations

It is important that all equipment beneath the raised floor be accurately located on as-built drawings. It is also a good idea to mark or label the floor tile where major components are located. For example, if there is a fan-powered box below the floor, it is helpful to mark the outline of the box on the floor tile and identify which tile is above the control box for that fan. This can be easily done with paint or marking pens before installing carpet. If the tile is left unfinished, other means of marking will have to be explored to maintain aesthetics.

The locations of the following equipment and components should be considered for documenting and marking:

- Fan-powered boxes
- Hydronic valves
- Fire/smoke dampers
- Air highway

- Balancing dampers
- Actuators
- Modular wiring junction boxes

6.2 BIM

If a building information model (BIM) was used during design and construction, it is possible to integrate this system with a facility management system. This can become a very useful tool for facility managers. The original model will have to be maintained throughout the life of the building and its systems. This is where all building components can be tracked for maintenance and other asset management activities.

References

ASHRAE. 2003. *Underfloor air distribution (UFAD) design guide.* pp. 96. Atlanta: ASHRAE.

ASHRAE. 2013. *UFAD guide: Design, construction, and operation of underfloor air distribution systems*. Atlanta: ASHRAE.

ASHRAE/USGBC/CIBSE. 2010. *Performance measurement protocols for commercial buildings*. Atlanta: ASHRAE, in collaboration with the Chartered Institute of Building Engineers, London, U.K., and U.S. Green Building Council, Washington, D.C.

Frontczak, M., S. Schiavon, J. Goins, E.A. Arens, H. Zhang, and P. Wargocki. 2012. Quantitative relationships between occupant satisfaction and satisfaction aspects of indoor environmental quality and building design. *Indoor Air* 22(2): 119–31

Jiang, Z., and Q. Chen. 2012. *Establishment of design procedures to predict room airflow requirements in partially mixed room air distribution systems* (RP-1522). ASHRAE Research Project, Final Report.

Hoyt, T., E. Arens, and H. Zhang. 2015. Extending air temperature setpoints: Simulated energy savings and design considerations for new and retrofit buildings. *Building and Environment* 88:89–96.

Lee, K.H., S. Schiavon, F.S. Bauman, and T. Webster. 2012. Thermal decay in underfloor air distribution (UFAD) systems: Fundamentals and influence on system performance. *Applied Energy* 91(1): 197–207.

Schiavon, S., and A.K. Melikov. 2008. Energy saving and improved comfort by increased air movement. *Energy and Buildings* 40(10):1954–60.

Schiavon, S., K.H. Lee, F.S. Bauman, and T. Webster. 2011. Simplified calculation method for design cooling loads in underfloor air distribution (UFAD) systems. *Energy and Buildings* 43(2-3): 517–528.

Schiavon, S., F.S. Bauman, B. Tully, and J. Rimmer. 2015. Chilled ceiling and displacement system: Laboratory study with high cooling load. *Science and Technology for the Built Environment* 21(7):944–56.

Webster, T., F. Bauman, S. Schiavon, D. Dickerhoff, D. Heinzerling, D. Troup, and D. Hill. 2011. Technical report on California state teachers retirement system building: UFAD performance and blinds study. CBE Report. Berkeley, CA: Center for the Built Environment, University of California Berkeley.